CONTENTS: TABLE OF CONTENTS

CONTENTS: WHAT IS THIS BOOK ABOUT?

STEM is an acronym for Science, Technology, Engineering, and Mathematics. Recent shifts in education have favored these subjects, primarily because we have a shortage of workforce in these particular areas, which is really quite sad for a number of reasons. These can be some of the most interesting things to study in school, provided they are taught in a fun, interesting, and hands-on fashion. They also lead to some of the best-paying technical jobs, too!

All of the labs within promote these 4 fields. You will find a strong emphasis on designing a project, testing it, measuring the results, and reflecting upon what worked and did not work. The projects are also labeled at the bottom with a series of categorical tags, so you can find similar projects to work on, allowing students to build upon prior knowledge gained in doing these projects! Of course, you can do them in any order you wish, but it can be fun to do a set of similar projects.

Since this is an educational volume, developed in my years teaching science in public schools, grading rubrics for each assignment are provided. There are some general suggestions and guidelines for each project, but it has deliberately been left without too much detail to allow the projects to be adapted to your classroom's individual needs.

CONTENTS: COPYRIGHT INFORMATION

CONTENTS: Alphabetical Mission Listing - Page 1

1. A Breathe of Fresh Air (**Air, Fliers, Height, Plastic Straws, Scavengers**)
2. Block and Tackle (**Glue, Paperclips, Popsicle Sticks, Spools, String, Weight, Wood**)
3. Blowguns (**Air, Distance, Plastic Straws, Throwers**)
4. Bottle Blasters (**Air, Bottles, Distance, Scavengers, Throwers**)
5. Crossbows (**Distance, Rubber Bands, Scavengers, Throwers**)
6. Egg Survivor VI - Flak Jackets (**Crashes, Eggs, Height, Scavengers, Survival**)
7. Eggs in a Basket (**Cups, Ping Pong Balls, Scavengers, Survival**)
8. Flippers (**Levers, Scavengers, Weight**)
9. Flushers (**Bottles, Cups, Scavengers, Water**)
10. Glue-ten Free (**Dead Lift, Glue, Materials Strength, String, Weight, Yarn**)
11. Golden Arches VI - Coins (**Arches, Coins, Height**)
12. Gone Fishin' (**Paperclips, Plastic Straws, Scavengers, Spools, String, Weight**)
13. Gummy Bridges (**Bridges, Gumdrops, Length, Toothpicks**)
14. Gummy Towers (**Gumdrops, Height, Toothpicks, Towers**)
15. Hand Mixer (**Gears, Paper Clips, Scavengers, Water**)
16. Homemade Orchestra I - Percussion (**Music, Scavengers, Sound, Time**)
17. Homemade Orchestra II - Strings (**Music, Scavengers, Sound, String, Time**)
18. Homemade Orchestra III - Wind (**Music, Scavengers, Sound, Time, Wind**)
19. Homemade Orchestra IV - Full Orchestra (**Music, Scavengers, Sound, Time, Wind**)
20. Jet-Puff Bridges (**Bridges, Marshmallows, Materials Strength, Pasta, Weight**)
21. King of Cups (**Cups, Height, Towers**)
22. Let's Get Cooking (**Chocolate, Foil, Heat, Melting, Scavengers, Sunlight, Time**)
23. Marshmallow Blaster (**Distance, Marshmallows, Scavengers, Throwers**)
24. Monster Truck Rally (**Cars, Crashes, Survival, Tracks**)
25. Peanut Tower (**Height, Packing Peanuts, Toothpicks, Towers**)

CONTENTS: Alphabetical Mission Listing - Page 2

26. Pinball Fever (**Clothespins, Marbles, Scavengers, Time**)
27. Running Uphill (**Cups, Scavengers, Time, Water**)
28. Salty Bridges (**Bridges, Glue, Length, Pretzels**)
29. Scavenger Bikes (**Bikes, Gears, Rubber Bands, Scavengers**)
30. Ski Jump (**Cardboard Tubes, Distance, Golf Balls, Marbles, Ping Pong Balls**)
31. Spaghetti Bridges (**Bridges, Glue, Materials Strength, Pasta, Weight**)
32. Starchy Goodness (**Glue, Height, String, Towers, Yarn**)
33. Sticky Planes (**Accuracy, Fliers, Paper, Tape, Velcro**)
34. Straw Rafts (**Boats, Buoyancy, Plastic Straws, Water, Weight**)
35. Stringy Situation (**Bridges, Glue, Materials Strength, String, Weight, Yarn**)
36. Suspension Bridges (**Bridges, Materials Strength, Scavengers, String, Weight**)
37. Target Practice (**Accuracy, Distance, Rubber Bands, Scavengers, Throwers**)
38. Throwing Money Away (**Clothespins, Coins, Distance, Throwers**)
39. Trebuchets (**Distance, Levers, Marbles, Scavengers, Throwers**)
40. Versus I - The Takedown (**Crashes, Scavengers, Survival, Throwers, Towers, Versus**)
41. Versus II - The Splashdown (**Boats, Crashes, Scavengers, Survival, Throwers, Versus, Water**)
42. Versus III - The Crashdown (**Crashes, Cups, Golf Balls, Scavengers, Survival, Throwers, Versus**)
43. Versus IV - The Rundown (**Cars, Crashes, Packing Peanuts, Toothpicks, Scavengers, Survival, Versus**)
44. Volume Up (**Popsicle Sticks, Rubber Bands, Volume)**
45. Washboard (**Scavengers, Sponges, Time, Water**)
46. Water Delivery (**Cups, Scavengers, Time, Water**)
47. Webbed Up (**Materials Strength, String, Weight, Yarn**)
48. What's for Breakfast? (**Cereal, Scavengers, Sorting**)
49. Wooden Cars (**Cars, Glue, Popsicle Sticks, Spools, Toothpicks, Wood**)
50. Wooden Railway (**Glue, Popsicle Sticks, Spools, Toothpicks, Tracks, Trains, Wood**)

CONTENTS: Mission Categories Listing

- Accuracy
- Air
- Arches
- Bikes
- Boats
- Bottles
- Bridges
- Buoyancy
- Cardboard Tubes
- Cars
- Cereal
- Chocolate
- Clothespins
- Coins
- Crashes
- Cups
- Dead Lift
- Distance
- Eggs
- Fliers
- Foil
- Gears
- Glue
- Golf Balls
- Gumdrops
- Heat
- Height
- Length
- Levers
- Marbles
- Marshmallows
- Materials Strength
- Melting
- Music
- Packing Peanuts
- Paper
- Paperclips
- Pasta
- Ping Pong Balls
- Plastic Straws
- Popsicle Sticks
- Pretzels
- Rubber Bands
- Scavengers
- Sorting
- Sound
- Sponges
- Spools
- String
- Sunlight
- Survival
- Tape
- Time
- Throwers
- Toothpicks
- Tracks
- Trains
- Towers
- Velcro
- Versus
- Volume
- Water
- Weight
- Wind
- Wood
- Yarn

CONTENTS: Categorical Mission Listing - Page 1

Accuracy

1. Sticky Planes
2. Target Practice

Air

1. A Breathe of Fresh Air
2. Blowguns
3. Bottle Blasters

Arches

1. Golden Arches VI

Bikes

1. Scavenger Bikes

Boats

1. Straw Rafts
2. Versus II

Bottles

1. Bottle Blasters
2. Flushers

Bridges

1. Gummy Bridges
2. Jet-Puff Bridges
3. Salty Bridges
4. Spaghetti Bridges
5. Stringy Situation
6. Suspension Bridges

Buoyancy

1. Straw Rafts

Cardboard Tubes

1. Ski Jump

Cars

1. Monster Truck Rally
2. Versus IV
3. Wooden Cars

Cereal

1. What's for Breakfast?

Chocolate

1. Let's Get Cooking

Clothespins

1. Pinball Fever
2. Throwing Money Away

Coins

1. Golden Arches VI
2. Throwing Money Away

Crashes

1. Egg Survivor VI
2. Monster Truck Rally
3. Versus I
4. Versus II
5. Versus III
6. Versus IV

Cups

1. Eggs in a Basket
2. Flushers
3. King of Cups
4. Running Uphill
5. Versus III
6. Water Delivery

CONTENTS: Categorical Mission Listing - Page 2

Dead Lift

1. Glue-ten Free

Distance

1. Blowguns
2. Bottle Blasters
3. Crossbows
4. Marshmallow Blaster
5. Ski Jump
6. Target Practice
7. Throwing Money Away
8. Trebuchets

Eggs

1. Egg Survivor VI

Fliers

1. A Breathe of Fresh Air
2. Sticky Planes

Foil

1. Let's Get Cooking

Gears

1. Hand Mixer
2. Scavenger Bikes

Glue

1. Block and Tackle
2. Glue-Ten Free
3. Salty Bridges
4. Spaghetti Bridges
5. Stringy Situation
6. Wooden Cars
7. Wooden Railway

Golf Balls

1. Ski Jump
2. Versus III

Gumdrops

1. Gummy Bridges
2. Gummy Towers

Heat

1. Let's Get Cooking

Height

1. A Breath of Fresh Air
2. Egg Survivor VI
3. Golden Arches VI
4. Gummy Towers
5. King of Cups
6. Peanut Tower
7. Starchy Goodness

Length

1. Gummy Bridges
2. Salty Bridges

Levers

1. Trebuchets
2. Flippers

Marbles

1. Pinball Fever
2. Ski Jump
3. Trebuchets

Marshmallows

1. Jet-Puff Bridges
2. Marshmallow Blaster

CONTENTS: Categorical Mission Listing - Page 3

Materials Strength

1. Glue-ten Free
2. Jet-Puff Bridges
3. Spaghetti Bridges
4. Stringy Situation
5. Suspension Bridges
6. Webbed Up

Melting

1. Let's Get Cooking

Music

1. Homemade Orchestra I
2. Homemade Orchestra II
3. Homemade Orchestra III
4. Homemade Orchestra IV

Packing Peanuts

1. Peanut Tower
2. Versus IV

Paper

1. Sticky Planes

Paperclips

1. Block and Tackle
2. Gone Fishin'
3. Hand Mixer

Pasta

1. Jet-Puff Bridges
2. Spaghetti Bridges

Ping Pong Balls

1. Eggs in a Basket
2. Ski Jump

Plastic Straws

1. A Breathe of Fresh Air
2. Blowguns
3. Gone Fishin'
4. Straw Rafts

Popsicle Sticks

1. Blck and Tackle
2. Volume Up
3. Wooden Cars
4. Wooden Railway

Pretzels

1. Salty Bridges

Rubber Bands

1. Crossbows
2. Scavenger Bikes
3. Target Practice
4. Volume Up

CONTENTS: Categorical Mission Listing - Page 4

Scavengers

1. A Breath of Fresh Air
2. Bottle Blasters
3. Crossbows
4. Egg Survivor VI
5. Eggs in a Basket
6. Flippers
7. Flushers
8. Gone Fishin'
9. Hand Mixer
10. Homemade Orchestra I
11. Homemade Orchestra II
12. Homemade Orchestra III
13. Homemade Orchestra IV
14. Let's Get Cooking
15. Marshmallow Blaster
16. Pinball Fever
17. Running Uphill
18. Scavenger Bikes
19. Suspension Bridges
20. Target Practice
21. Trebuchets
22. Versus I
23. Versus II
24. Versus III
25. Versus IV
26. Washboard
27. Water Delivery
28. What's for Breakfast?

Sorting

1. What's for Breakfast?

Sound

1. Homemade Orchestra I
2. Homemade Orchestra II
3. Homemade Orchestra III
4. Homemade Orchestra IV

Sponges

1. Washboard

Spools

1. Block and Tackle
2. Gone Fishin'
3. Wooden Cars
4. Wooden Railway

String

1. Block and Tackle
2. Glue-Ten Free
3. Gone Fishin'
4. Homemade Orchestra II
5. Starchy Goodness
6. Stringy Situation
7. Suspension Bridges
8. Webbed Up

Sunlight

1. Let's Get Cooking

Survival

1. Egg Survivor VI
2. Eggs in a Basket
3. Monster Truck Rally
4. Versus I
5. Versus II
6. Versus III
7. Versus IV

Tape

1. Sticky Planes

CONTENTS: Categorical Mission Listing - Page 5

Time

1. Homemade Orchestra I
2. Homemade Orchestra II
3. Homemade Orchestra III
4. Homemade Orchestra IV
5. Let's Get Cooking
6. Pinball Fever
7. Running Uphill
8. Washboard
9. Water Delivery

Throwers

1. Blowguns
2. Bottle Blasters
3. Crossbows
4. Marshmallow Blaster
5. Target Practice
6. Throwing Money Away
7. Trebuchets
8. Versus I
9. Versus II
10. Versus III

Toothpicks

1. Gummy Bridges
2. Gummy Towers
3. Peanut Tower
4. Versus IV
5. Wooden Cars
6. Wooden Railway

Tracks

1. Monster Truck Rally
2. Wooden Railway

Trains

1. Wooden Railway

Towers

1. Gummy Towers
2. King of Cups
3. Peanut Tower
4. Starchy Goodness
5. Versus I

Velcro

1. Sticky Planes

Versus

1. Versus I
2. Versus II
3. Versus III
4. Versus IV

Volume

1. Volume Up

Water

1. Flushers
2. Hand Mixer
3. Running Uphill
4. Straw Rafts
5. Versus II
6. Washboard
7. Water Delivery Service

CONTENTS: Categorical Mission Listing - Page 6

Weight

1. Block and Tackle
2. Flippers
3. Glue-Ten Free
4. Jet-Puff Bridges
5. Spaghetti Bridges
6. Straw Rafts
7. Suspension Bridges
8. Webbed Up

Wind

1. Homemade Orchestra III
2. Homemade Orchestra IV

Wood

1. Block and Tackle
2. Wooden Cars
3. Wooden Railway

Yarn

1. Glue-Ten Free
2. Starchy Goodness
3. Stringy Situation
4. Webbed Up

CONTENTS: 50 NEW STEM LABS

Each of the new science labs in **50** NEW **STEM LABS** has the following:

- A snappy **Title**
- A **Brief Description** of the task to be completed
- General **Mission Rules**, suggestions, limitations, and requirements of the task
- **Grading Rubrics** for a Quiz and a Test Grade
- A small **note space** for any changes or adaptations required
- **Category Tags** at the bottom to help you find similar projects

MISSION: A Breath of Fresh Air

BRIEF: You and your team have been selected to make a rocket that can fly as high as possible on a single breath of air.

MISSION RULES:

1. You will design a rocket of any dimensions you wish.
2. You will work with one to two partners. Teams may not be of more than 3 people.
3. You may use whatever allowed materials you can find at school or home in your project.
4. Your device must be have a plastic straw tube inside of it. A second, smaller straw will be inserted into the first, and a breath of air will be blown into it to give it lift.

TEACHER'S NOTES: It is suggested that you use two different sizes of straws. The larger one should be in the rocket. The smaller one should be put inside the first, and the students will blow into the end of it to give the device lift.

Additionally, it might help if students were to all sit or lie down, so that each test was launched from a specific height. Graduated marks along a wall or measuring tapes along the wall can also help gauge heights.

QUIZ GRADE:

Create a blueprint design for your ideas

- Sketch 25%
- Sketch is labeled 25%
- Explanation of strategies 25%
- Conclusions and reflections based on your results 25%

TEST GRADE:

Your completed design and the results of the test.

- Project Completed = 50%
- 50% of your grade depends on how high your project flies compared to others.
- Top scores get +50%, and those following get +40%, +30%, or +20%.

NOTES:

CATEGORIES: Air, Fliers, Height, Plastic Straws, Scavengers

MISSION: Block and Tackle

BRIEF: You and your team have been selected to make a wooden block and tackle.

MISSION RULES:

1. You will design a block and tackle using just dowel rods, thread spools, popsicle sticks, and glue.
2. The spools must be able to spin and turn with the movement of the string or line.
3. You will work with one or two partners. Teams may be of no more than 3 people.
4. Your device must have paperclips to anchor the top and to hook weights and strings to.
5. Weight will be gradually added to the system, until it breaks or is near breaking. Holding more weight means a better score.

TEACHER'S NOTES: Wax paper is suggested as a non-sticky surface for glued pieces to dry upon.

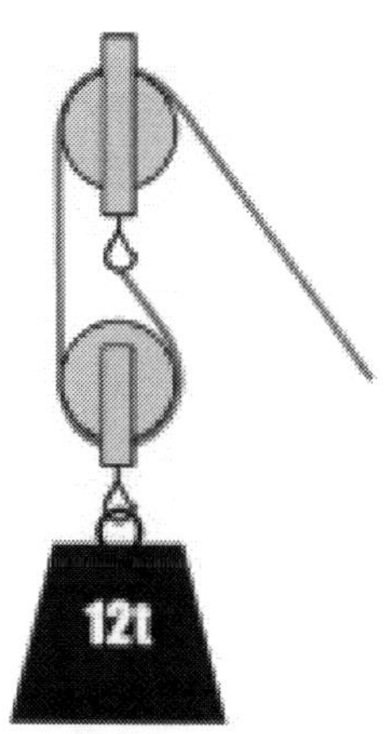

QUIZ GRADE:

A blueprint design of your idea

- Sketch 25%
- Sketch is labeled 25%
- Explanation of strategy 25%
- Conclusions and reflections based on your results 25%

TEST GRADE:

Your completed design and the results of the test.

- Project Completed = 50%
- 50% of your grade depends on how much weight your device holds before breaking.
- *NOTE: The car that moves the farthest gets an automatic 100%*

NOTES:

CATEGORIES: Glue, Paperclips, Popsicle Sticks, Spools, String, Weight, Wood

MISSION: Blowguns

BRIEF: You and your team have been selected to make a projectile that can be shot as far as possible with a plastic straw.

MISSION RULES:

1. You will design a set of **harmless** projectiles that shoot as far as possible by blowing into a plastic straw.
2. Your projectiles/devices may be of any dimensions, but they must have a receiver that allows it to fit in or around the end of the straw.
3. You may build your device from any approved materials found at school or at home.
4. You will work with one or two partners. Teams may be of no more than 3 people.
5. Up to three tests will be made. Distance will be measured at the first contact with the ground.

TEACHER'S NOTES: After 3 tests, honors can be given for the best average and/or the highest single shot.

PLEASE NOTE: Take precautions not to inhale any projects or to hit anyone in the eyes.

QUIZ GRADE:

A blueprint design of your idea

- Sketch 25%
- Sketch is labeled 25%
- Explanation of strategy 25%
- Conclusions and reflections based on your results 25%

TEST GRADE:

Your completed design and the results of the test.

- Project Completed = 50%
- 50% of your grade depends on how far your project sends the projectile.
- *NOTE: The best project gets an automatic 100%.*

NOTES:

CATEGORIES: Air, Distance, Plastic Straws, Throwers

MISSION: Bottle Blasters

BRIEF: You and your team have been selected to make a device that flies as far as possible by squeezing an empty plastic bottle.

MISSION RULES:

1. You will design a **harmless** flying device that is propelled by squeezing (or stomping on) an empty plastic bottle.

2. Your devices may be of any dimensions, but they must have a receiver that allows it to fit in or around the cap end of the plastic bottle.

3. You may build your device from any approved materials found at school or at home.

4. You will work with one or two partners. Teams may be of no more than 3 people.

5. Up to three tests will be made. Distance will be measured at the first contact with the ground.

TEACHER'S NOTES: After 3 tests, honors can be given for the best average and/or the highest single shot.

Please take eye safety precautions.

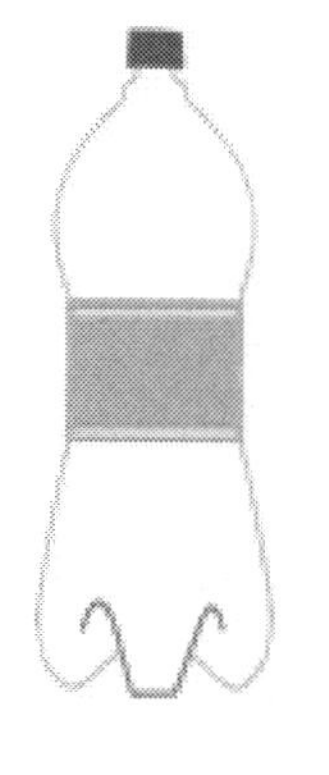

QUIZ GRADE:

A blueprint design of your idea

- Sketch 25%
- Sketch is labeled 25%
- Explanation of strategy 25%
- Conclusions and reflections based on your results 25%

TEST GRADE:

Your completed design and the results of the test.

- Project Completed = 50%
- 50% of your grade depends on how far your project flies
- *NOTE: The best project gets an automatic 100%.*

NOTES:

CATEGORIES: Air, Bottles, Distance, Scavengers, Throwers

MISSION: Crossbows

BRIEF: You and your team have been selected to make a device that shoots an arrow as far as possible.

MISSION RULES:

1. You will design a device that shoots a blunt or **harmless** arrow as far as possible using rubber bands as your major method of propulsion.
2. Your device may be of any dimensions under 9 inches in any one direction.
3. You may build your device from any approved materials found at school or at home.
4. The device must be freestanding and not attached to any surface.
5. You will work with one or two partners. Teams may be of no more than 3 people.
6. Up to three tests will be made. Distance will be measured at the first contact with the ground.

TEACHER'S NOTES: After 3 tests, honors can be given for the best average and/or the highest single shot.

You can also test for accuracy.

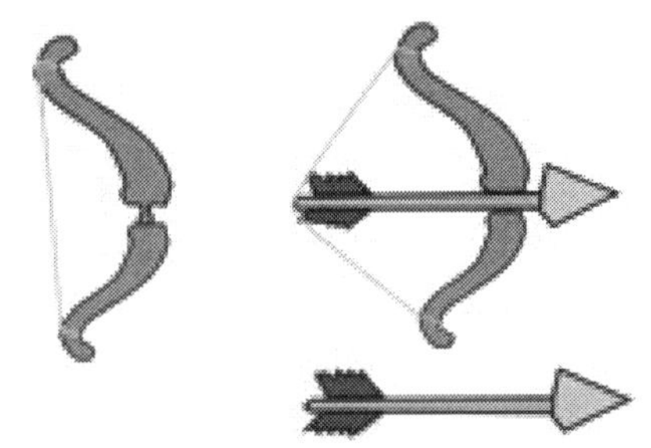

QUIZ GRADE:

A blueprint design of your idea

- Sketch 25%
- Sketch is labeled 25%
- Explanation of strategy 25%
- Conclusions and reflections based on your results 25%

TEST GRADE:

Your completed design and the results of the test.

- Project Completed = 50%
- 50% of your grade depends on how far your project sends the arrow.
- *NOTE: The best project gets an automatic 100%.*

NOTES:

CATEGORIES: Distance, Rubber Bands, Scavengers, Throwers

MISSION: Eggs in a Basket

BRIEF: You and your team have been selected to make a device that allows a ping pong ball in a cup to land on the floor safely without the ball falling out of the cup.

MISSION RULES:

1. You will design a device that allows a ping pong ball in a cup to land on the floor safely without the ball falling out of the cup.
2. Your device may be of any dimensions under 18 inches in any one direction.
3. You may build your device from any approved materials found at school or at home.
4. Your cup may not be covered at all. Only hooks/line may be attached around the rim of the cup to allow it to be hooked to your project.
5. You will work with one or two partners. Teams may be of no more than 3 people.
6. Up to three tests will be made. Projects will be dropped from a height of no less than 6 feet.

TEACHER'S NOTES: Standing on top of a chair/table or step ladder is advised. Dropping from farther up may give certain projects a chance to deploy.

Substitute a marble if ping pong balls are not available.

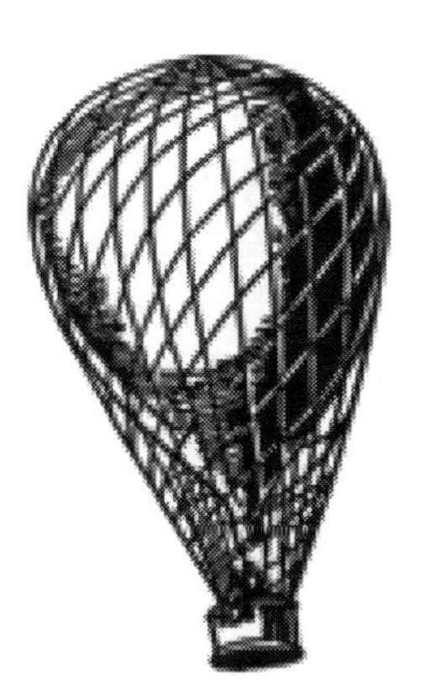

QUIZ GRADE:

A blueprint design of your idea

- Sketch 25%
- Sketch is labeled 25%
- Explanation of strategy 25%
- Conclusions and reflections based on your results 25%

TEST GRADE:

Your completed design and the results of the test.

- Project Completed = 50%
- 50% of your grade depends on test results
    - Survive 3 landings = 50%
    - Survive 2 landings = 35%
    - Survive 1 landing = 20%
    - Survive 0 landings = 0%

NOTES:

CATEGORIES: Cups, Ping Pong Balls, Scavengers, Survival

MISSION: Flippers

BRIEF: You and your team have been selected to make a device that can flip things over.

MISSION RULES:

1. You will design a device that can flip over as much weight as possible.
2. Your device should not be over 12 inches in any dimension.
3. You will work with one to two partners. Teams may not be of more than 3 people.
4. You may use any approved materials from home or school to make your device.
5. Your device must be free-standing and movable. It cannot be attached to any surface.
6. Your device should have some sort of skid that slides under the books or weight to be flipped.
7. Your device should have a lever or activator that activates your machine's flipping action.

TEACHER'S NOTES: It is suggested that you use paperback books for flipping. The devices will be activated and should flip the books over.

QUIZ GRADE:

Create a blueprint design for your ideas

- Sketch 25%
- Sketch is labeled 25%
- Explanation of strategies 25%
- Conclusions and reflections based on your results 25%

TEST GRADE:

Your completed design and the results of the test.

- Project Completed = 50%
- 50% of your grade depends on how much weight your project can flip over compared to others.
- Top scores get +50%, and those following get +40%, +30%, or +20%.

NOTES:

CATEGORIES: Levers, Scavengers, Weight

MISSION: Flushers

BRIEF: You and your team have been selected to make a working toilet.

MISSION RULES:

1. You will design a device that can flush dirty water into a holding tank, replacing it with fresh water.
2. Your device should not be over 18 inches in any dimension.
3. You will work with two to three partners. Teams may not be of more than 4 people.
4. You may use any approved materials from home or school to make your device.
5. Your device must be free-standing and movable. It cannot be attached to any surface.
6. Your device should have a lever or activator that activates your machine's flushing action.
7. Your device should have 3 separate holding areas: the fresh water tank, the bowl, and the dirty water tank. Water should move between then upon activating your machine.

TEACHER'S NOTES: It is suggested that you use food coloring or plastic beads to simulate the 'dirty' water and to determine if it is being cleaned as it is drained and replaced.

QUIZ GRADE:

Create a blueprint design for your ideas

- Sketch 25%
- Sketch is labeled 25%
- Explanation of strategies 25%
- Conclusions and reflections based on your results 25%

TEST GRADE:

Your completed design and the results of the test.

- Project Completed = 50%
- 50% of your grade depends on how well your project works.
    - Project replaces water +25%
    - Project cleans water +15%
    - Project can be flushed twice +10%

NOTES:

CATEGORIES: Bottles, Cups, Scavengers, Water

MISSION: Egg Survivor VI - Flak Jackets

BRIEF: You and your team have been selected to make a new EPD (egg protection device) to protect an egg in a fall. This one must only cover the egg like a jacket, and may not be more than 1 inch thick.

MISSION RULES:

1. You will design a vehicle of dimensions as small as possible that can survive as high of a drop as possible.
2. All materials must be in direct contact with the eggshell. They may not be more than 1" thick, either.
3. No matter how the egg is coated or covered, there must be a way to open it up or look to see if the egg is still intact.
4. You may use any materials you want, provided you can scrounge them up, buy them, or find them.
5. You may test at home. In fact, you're encouraged to test at home! Assembly and design may also take place at school, but time is limited.
6. Teams may be of no more than 3 people.

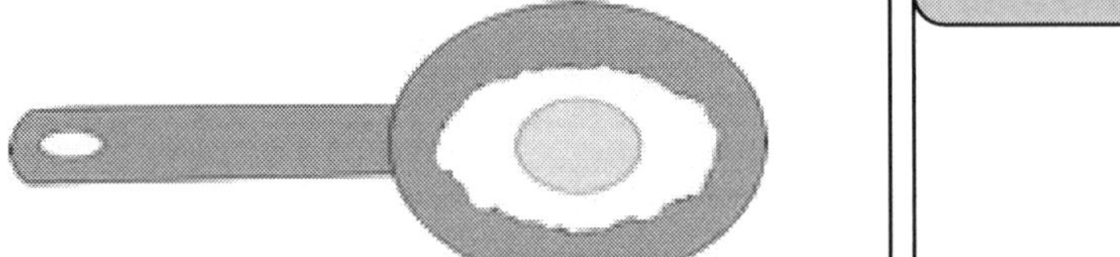

QUIZ GRADE:

Create a blueprint designs for your ideas

- Labeled Sketch 25%
- Materials List 25%
- Explanation of strategies 25%
- Conclusions and reflections based on your results 25%

TEST GRADE:

Your completed design and the results of the test.

- Completed Design = 50% (penalties assessed if it does not follow the rules)
- 50% of your score depends on how high of a drop it survives.
- Points will be given in 10% increments for each drop it survives: 10, 20, 30, 40, and finally 50%.
- The best project automatically gets 100%

NOTES:

CATEGORIES: Crashes, Eggs, Height, Scavengers, Survival

MISSION: Glue-ten Free

BRIEF: You and your team have been selected to make the strongest possible structure out of glue and string or yarn.

MISSION RULES:

1. You will design a device that is at least 4 inches in all three dimensions, which can hold up as much weight as possible without collapsing.
2. Your teacher will determine the maximum amount of materials you may use in your project.
3. You will work with one to two partners. Teams may not be of more than 3 people.
4. You may use only string/yarn and glue for your construction.
5. Your device must be free-standing and movable. It cannot be attached to any surface.

TEACHER'S NOTES: It is suggested that you use wax paper to set up 'cables' of thread or yarn. Glue will help them set into a specific shape when it dries. Then they can be layered and attached to form the structure.

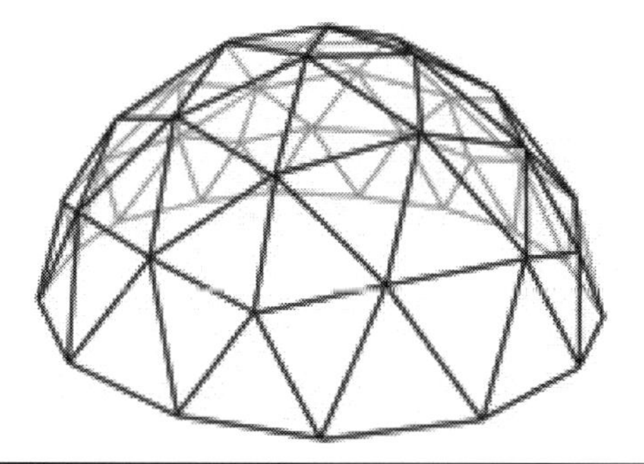

QUIZ GRADE:

Create a blueprint design for your ideas

- Sketch 25%
- Sketch is labeled 25%
- Explanation of strategies 25%
- Conclusions and reflections based on your results 25%

TEST GRADE:

Your completed design and the results of the test.

- Project Completed = 50%
- 50% of your grade depends on how much weight your project can hold up compared to others.
- Top scores get +50%, and those following get +40%, +30%, or +20%.

NOTES:

CATEGORIES: Dead Lift, Glue, Materials Strength, String, Weight, Yarn

MISSION: Golden Arches VI - Coins

BRIEF: You and your team have been selected to design an arch that is as tall and wide as possible from just coins.

MISSION RULES:

1. You will design an arch that is as tall and as wide as possible.
2. Your finished project must be built of only coins.
3. Your project must be completely freestanding and may not be attached to the floor or a table surface.
4. Your teacher will determine your materials limit for the arch.
5. You will work with one or two partners. Teams may be of no more than 3 people.
6. While a perfect curve is hard to attain, some semblance of an arch must be created. A simple V or single-pitched roof is not acceptable.

QUIZ GRADE:

Create a blueprint design for your ideas

- Sketch 25%
- Sketch is labeled 25%
- Explanation of strategies 25%
- Conclusions and reflections based on your results 25%

TEST GRADE:

Your completed design and the results of the test.

- Project Completed = 50%
- 50% of your project's score depends on the height and width of your project as compared to other projects.
- A formula of width X height will be used. Width is at the widest part near the base, and height is as the peak of the arch.
- *NOTE: The arch with the highest score will automatically get 100%*

NOTES:

CATEGORIES: Arches, Coins, Height

MISSION: Gone Fishin'

BRIEF: You and your team have been selected to make a working fishing pole.

MISSION RULES:

1. You will design a working fishing pole. It will wind up cord when a hand crank is turned.
2. The fishing pole and reel may be built from any approved materials found at home or at school. Suggested materials are: plastic straws, cardboard, rubber bands, paperclips, glue, tape, thread spools, etc...
3. You will work with one or two partners. Teams may be of no more than 3 people.
4. Your device's reel should be less than 6 inches in all dimensions. The pole may be up to 3 feet long.
5. The pole must have eyelets to thread the line through, and the line must have a paperclip hook on the end.
6. The pole will be strength-tested to see how large of a 'fish' or weight it can pull in.

TEACHER'S NOTES: Suggested 'fish' are graduated weights with eyelets on them. Toy fish of varying weights can also be used.

Reels don't have to cast, but they could. They should at least allow the line to be pulled back out.

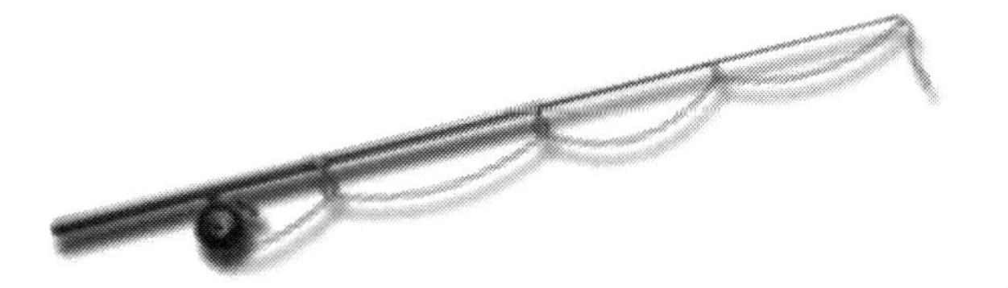

QUIZ GRADE:

A research paper on fishing reels.

- 2-3 pictures of fishing reels 25%
- A labeled concept idea based on your fishing reel pictures, including what materials you hope to use for each piece 50%
- Conclusions and reflections based on your results 25%

TEST GRADE:

Your completed design and the results of the test.

- Project Completed = 50%
- Reel works = 20%
- 30% of your grade depends on how much weight your reel and pole can handle compared to other projects.

NOTE: The best project gets 100%

NOTES:

CATEGORIES: Paperclips, Plastic Straws, Scavengers, Spools, String, Weight

MISSION: Gummy Bridges

BRIEF: You and your team have been selected to make as long of a bridge as possible using only gumdrops and toothpicks.

MISSION RULES:

1. You will research bridges and get ideas for a concept for your bridge design.
2. Your bridge must be as long as possible using the provided materials.
3. You will work with one or two partners. Teams may not be of more than 3 people.
4. You must only use gumdrops and toothpicks for your project. You teacher will determine how many materials you may use.
5. The bridge must not collapse for at least 10 seconds when it is being tested.

QUIZ GRADE:

A research paper on bridges.

- 2-4 pictures of bridges 25%
- A concept idea based on your bridge pictures 50%
- Conclusions and reflections based on your results 25%

TEST GRADE:

Your completed design and the results of the test.

- Project Completed = 50%
- 50% of your grade depends on how long your project is compared to the other group's projects. The projects that do best will get more points.

NOTES:

CATEGORIES: Bridges, Gumdrops, Length, Toothpicks

MISSION: Gummy Bridges

BRIEF: You and your team have been selected to make as tall of a tower as possible using only gumdrops and toothpicks.

MISSION RULES:

1. You will research towers and get ideas for a concept for your tower design.
2. Your tower must be as tall as possible using the provided materials.
3. You will work with one or two partners. Teams may not be of more than 3 people.
4. You must only use gumdrops and toothpicks for your project. You teacher will determine how many materials you may use.
5. The tower must be freestanding and may not be attached to any surface.
6. The tower must not collapse for at least 10 seconds when it is being tested.

QUIZ GRADE:

A research paper on towers.

- 2-4 pictures of towers 25%
- A concept idea based on your tower pictures 50%
- Conclusions and reflections based on your results 25%

TEST GRADE:

Your completed design and the results of the test.

- Project Completed = 50%
- 50% of your grade depends on how tall your project is compared to the other group's projects. The projects that do best will get more points.

NOTES:

CATEGORIES: Gumdrops, Height, Toothpicks, Towers

MISSION: Hand Mixer

BRIEF: You and your team have been selected to make a working hand mixer.

MISSION RULES:

1. You will design a working hand mixer. It will use gears or some sort of drive to turn the mixer(s) when you turn the hand crank.
2. The mixer may be built from any approved materials found at home or at school. Suggested materials are: plastic straws, cardboard, rubber bands, paperclips, glue, tape, etc...
3. You will work with one or two partners. Teams may be of no more than 3 people.
4. Your device may be of any dimensions less than 12 inches.

TEACHER'S NOTES: Style and Function awards are a cool option.

QUIZ GRADE:

A research paper on mixers.

- 2-3 pictures of hand crank mixers 25%
- A labeled concept idea based on your mixer pictures, including what materials you hope to use for each piece 50%
- Conclusions and reflections based on your results 25%

TEST GRADE:

Your completed design and the results of the test.

- Project Completed = 50%
- 50% of your grade depends on how well your project works to stir a bowl of water.
  - Works: Well +50%, Okay 30%, Hardly at all, 10%
  - More than 1 set of mixer tines: +20%

NOTES:

CATEGORIES: Gears, Paper Clips, Scavengers, Water

MISSION: Homemade Orchestra I - Percussion

BRIEF: You and your team have been selected to make a custom percussion instrument from scavenged materials.

MISSION RULES:

1. You will research percussion instruments to get ideas for a concept for your design.
2. Your design may be of any reasonable dimensions.
3. You will work with two or three partners. Teams may not be of more than 4 people.
4. You may use any approved scavenged materials from home and school to build your device.
5. Your device must have a minimum of 3 different audible tones. It can be played by hand or with some sort of stick.
6. Your team must compose and perform a song at least 30 seconds in length with your instrument. Practice makes perfect!

QUIZ GRADE:

A research paper on percussion instruments.

- 2-4 pictures of percussion instruments 25%
- A concept idea for your percussion instruments 50%
- Conclusions and reflections based on your results 25%

TEST GRADE:

Your completed design and the results of the test.

- Project Completed = 60%
- 40% of your grade depends on your instrument's performance. If the instrument breaks during the performance, it will cost points.
- Writing the music on music sheet notes is a BONUS!

NOTES:

CATEGORIES: Music, Scavengers, Sound, Time

MISSION: Homemade Orchestra II - Strings

BRIEF: You and your team have been selected to make a custom stringed instrument from scavenged materials.

MISSION RULES:

1. You will research stringed instruments to get ideas for a concept for your design.
2. Your design may be of any reasonable dimensions.
3. You will work with two or three partners. Teams may not be of more than 4 people.
4. You may use any approved scavenged materials from home and school to build your device.
5. Your device must have a minimum of 3 strings that give different audible tones. It can be plucked, strummed, or it may require a custom-made pick or bow to play it.
6. Your team must compose and perform a song at least 30 seconds in length with your instrument. Practice makes perfect!

QUIZ GRADE:

A research paper on stringed instruments.

- 2-4 pictures of stringed instruments 25%
- A concept idea for your stringed instruments 50%
- Conclusions and reflections based on your results 25%

TEST GRADE:

Your completed design and the results of the test.

- Project Completed = 60%
- 40% of your grade depends on your instrument's performance. If the instrument breaks during the performance, it will cost points.
- Writing the music on music sheet notes is a BONUS!

NOTES:

CATEGORIES: Music, Scavengers, Sound, String, Time

MISSION: Homemade Orchestra III - Wind

BRIEF: You and your team have been selected to make a custom wind (brass or woodwind) instrument from scavenged materials.

MISSION RULES:

1. You will research wind instruments to get ideas for a concept for your design.
2. Your design may be of any reasonable dimensions.
3. You will work with one or two partners. Teams may not be of more than 3 people.
4. You may use any approved scavenged materials from home and school to build your device.
5. Your device must have a minimum of 3 different audible tones.
6. Your team must compose and perform a song at least 30 seconds in length with your instrument. Practice makes perfect!

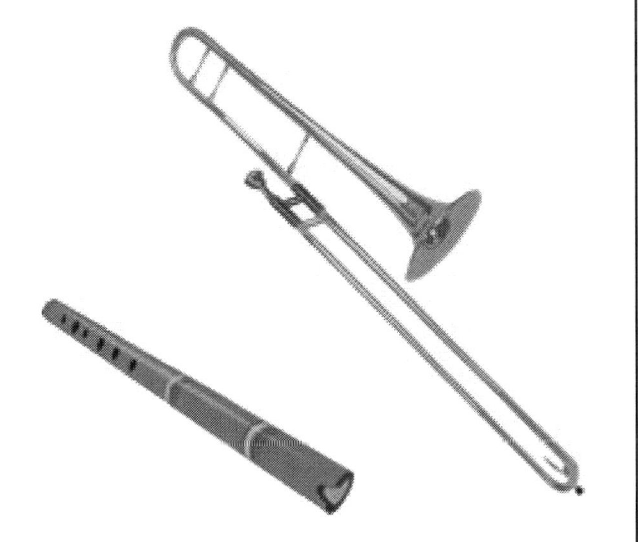

QUIZ GRADE:

A research paper on wind instruments (including brass and woodwind).

- 2-4 pictures of wind instruments 25%
- A concept idea for your wind instruments 50%
- Conclusions and reflections based on your results 25%

TEST GRADE:

Your completed design and the results of the test.

- Project Completed = 60%
- 40% of your grade depends on your instrument's performance. If the instrument breaks during the performance, it will cost points.
- Writing the music on music sheet notes is a BONUS!

NOTES:

CATEGORIES: Music, Scavengers, Sound, Time, Wind

MISSION: Homemade Orchestra IV - Full Orchestra

BRIEF: You and your team have been selected to make at least 3 instruments of different families and compose a tune together.

MISSION RULES:

1. You will research instruments to get ideas for a concept for your designs.
2. Your design may be of any reasonable dimensions.
3. You will work with three to five partners. Teams may not be of more than 6 people.
4. You may use any approved scavenged materials from home and school to build your device.
5. Each device must have a minimum of 3 different audible tones.
6. Your team must compose and perform a song at least 60 seconds in length with your instruments. Practice makes perfect!

QUIZ GRADE:

A research paper on typical instruments in an orchestra.

- 5-10 pictures of wind instruments 25%
- Concept ideas for each of your instruments 50%
- Conclusions and reflections based on your results 25%

TEST GRADE:

Your completed design and the results of the test.

- Project Completed = 60%
- 40% of your grade depends on your band's performance. If the instruments break during the performance, it will cost points.
- Writing the music on music sheet notes is a BONUS!

NOTES:

CATEGORIES: Music, Scavengers, Sound, Time, Wind

MISSION: Jet-Puff Bridges

BRIEF: You and your team have been selected to make as strong of a bridge as possible from pasta and marshmallows!

MISSION RULES:

1. You will research bridges and get ideas for a concept for your bridge design.
2. Your bridge must be 18 inches long, between 2 and 6 inches wide, and 2 to 12 inches tall. If you are outside these measurements by more than 1/2 inch, you will be penalized.
3. You will work with two or three partners. Teams may not be of more than 4 people.
4. You must only use pasta and marshmallows for your project. You teacher will determine how many materials you may use.
5. The bridge must have a place in the center of it where a tray can be attached to hold weight. Projects that hold more weight score better.

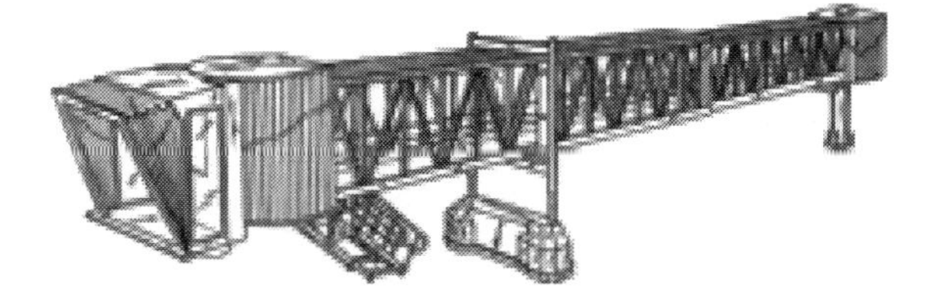

QUIZ GRADE:

A research paper on bridges.

- 2-4 pictures of bridges 25%
- A concept idea based on your bridge pictures 50%
- Conclusions and reflections based on your results 25%

TEST GRADE:

Your completed design and the results of the test.

- Project Completed = 50%
- 50% of your grade depends on how much weight your project holds compared to the other group's projects. The projects that do best will get more points.
- *NOTE: There is a -5% penalty for every 1/2 inch your project is out of the specifications.*

NOTES:

CATEGORIES: Bridges, Marshmallows, Materials Strength, Pasta, Weight

MISSION: King of Cups

BRIEF: You and your team have been selected to make a the tallest tower possible from only plastic cups!

MISSION RULES:

1. You will design a tower.
2. Your device may be any dimensions, but it must be as tall as possible.
3. You will work with one or two partners. Teams may not be of more than 3 people.
4. You may use only plastic cups to build your project. Your teacher will determine your maximum amount of materials.
5. The tower may not be braced against or attached to any other objects, except for whatever table surface it is built on. Otherwise, it must be completely freestanding.
6. Your tower must stand on its own for at least 10 seconds while it is being measured.

TEACHER'S NOTES: A set number, like 10, 12, or 15 cups is suggested.

QUIZ GRADE:

A blueprint design of your idea

- Sketch 25%
- Sketch is labeled 25%
- Explanation of strategy 25%
- Conclusions and reflections based on your results 25%

TEST GRADE:

Your completed design and the results of the test.

- Project Completed = 50%
- 50% of your grade depends on how tall your project is compared to the other group's projects. The projects that do best will get more points.

NOTES:

CATEGORIES: Cups, Height, Towers

MISSION: Let's Get Cooking

BRIEF: You and your team have been selected to make a solar oven.

MISSION RULES:

1. You will design an oven.
2. Your device may be any dimensions less than 12 inches in any one direction.
3. You will work with one or two partners. Teams may not be of more than 3 people.
4. You may use only foil and other approved materials found at home or school to build your oven.
5. The oven must melt pieces of chocolate as completely and quickly as possible. A paperclip or probe will be used to determine how melted the chocolate is, if there is any question.

TEACHER'S NOTES: small pieces of milk chocolate (3 or more) should be spread around the baking surface in full sun to determine which ovens work best.

QUIZ GRADE:

A blueprint design of your idea

- Sketch 25%
- Sketch is labeled 25%
- Explanation of strategy 25%
- Conclusions and reflections based on your results 25%

TEST GRADE:

Your completed design and the results of the test.

- Project Completed = 50%
- 50% of your grade depends on how fast and completely your project melts the chocolate. Better projects score more points.

NOTES:

CATEGORIES: Chocolate, Foil, Heat, Melting, Scavengers, Sunlight, Time

MISSION: Marshmallow Blaster

BRIEF: You and your team have been selected to make a cannon to shoot a marshmallow through a tube as far as possible.

MISSION RULES:

1. You will design a device that launches a marshmallow through a tube as far as possible.
2. Your primary building materials should be a cardboard tube or some pipe, plus whatever scavenged materials you can find to help propel the marshmallow.
3. You will work with one or two partners. Teams may be of no more than 3 people.
4. Up to three tests will be made.

TEACHER'S NOTES: After 3 tests, honors can be given for the best average and/or the longest single shot.

QUIZ GRADE:

A blueprint design of your idea

- Sketch 25%
- Sketch is labeled 25%
- Explanation of strategy 25%
- Conclusions and reflections based on your results 25%

TEST GRADE:

Your completed design and the results of the test.

- Project Completed = 50%
- 50% of your grade depends on how far your project sends the marshmallows.
- *NOTE: The best project gets an automatic 100%.*

NOTES:

CATEGORIES: Distance, Marshmallows, Scavengers, Throwers

MISSION: Monster Truck Rally

BRIEF: You and your team have been selected to make the strongest monster trucks possible to crash and smash your opposition.

MISSION RULES:

1. You will design a free-rolling monster truck. It may have additional propulsion from rubber band, balloons, or other homemade type drive engines.

2. The car may be no more than 6 inches tall or wide and 9 inches long.

3. Your design must have a spot for an army man or other small driver character. Your driver may not be completely encaged or taped/glued down. There must be an adequate chance of being thrown from the car.

4. You will work with one or two partners. Teams may not be of more than 3 people.

5. You may use any approved materials to build your project. Your teacher will determine your maximum amount of materials.

TEACHER'S NOTES: A dual track design is required. Two wooden boards with shallow ledges on the sides are suggested. They should be arranged in a slight V-formation, so the cars roll toward each other and impact.

Determine your own rules and scoring method and/ or a bracket system to determine final winner.

QUIZ GRADE:

A blueprint design of your idea

- Sketch 25%
- Sketch is labeled 25%
- Explanation of strategy 25%
- Conclusions and reflections based on your results 25%

TEST GRADE:

Your completed design and the results of the test.

- Project Completed = 50%
- 50% of your grade depends on how much well your project survives compared to others.
  - VICTORY CONDITIONS: a car drives over another, flips a card over, causes the other car to lose its driver, or disables a car due to damage.
  - ELIMINATION: defeated in the ways listed above, or if your car falls off of the track, does not make it down the track, flips over on its own, or is unable to continue due to self-inflicted damage.

CATEGORIES: Cars, Crashes, Survival

MISSION: Peanut Tower

BRIEF: You and your team have been selected to make as tall of a tower as possible using only foam packing peanuts and toothpicks.

MISSION RULES:

1. You will research towers and get ideas for a concept for your tower design.
2. Your tower must be as tall as possible using the provided materials.
3. You will work with one or two partners. Teams may not be of more than 3 people.
4. You must only use packing peanuts and toothpicks for your project. You teacher will determine how many materials you may use.
5. The tower must be freestanding and may not be attached to any surface.
6. The tower must not collapse for at least 10 seconds when it is being tested.

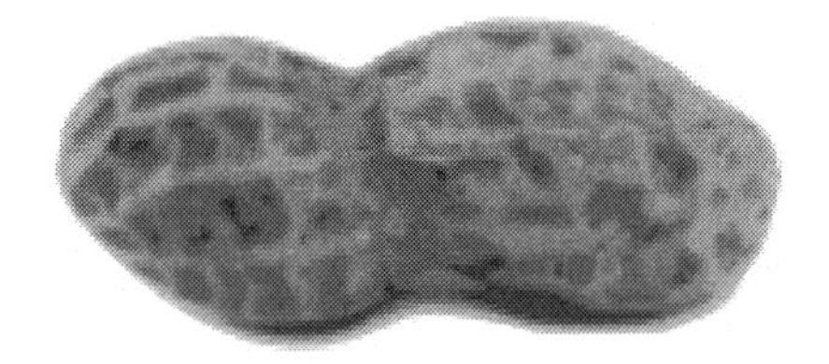

QUIZ GRADE:

A research paper on towers.

- 2-4 pictures of towers 25%
- A concept idea based on your tower pictures 50%
- Conclusions and reflections based on your results 25%

TEST GRADE:

Your completed design and the results of the test.

- Project Completed = 50%
- 50% of your grade depends on how tall your project is compared to the other group's projects. The projects that do best will get more points.

NOTES:

CATEGORIES: Height, Packing Peanuts, Toothpicks, Towers

MISSION: Pinball Fever

BRIEF: You and your team have been selected to make a pinball machine for a marble, which will take as long as possible for the marble to return to the launch.

MISSION RULES:

1. You will research pinball machines and get ideas for a concept for your design.
2. Your design may be of any dimensions less than 3 feet in any one direction.
3. You will work with two or three partners. Teams may not be of more than 4 people.
4. You may use any approved scavenged materials from home and school to build your device.
5. Your device will use a clothespin to launch the marble.
6. The marble will go through a series of tricks or structures to delay the return of the marble to the start point. The longer it takes to return, the better.

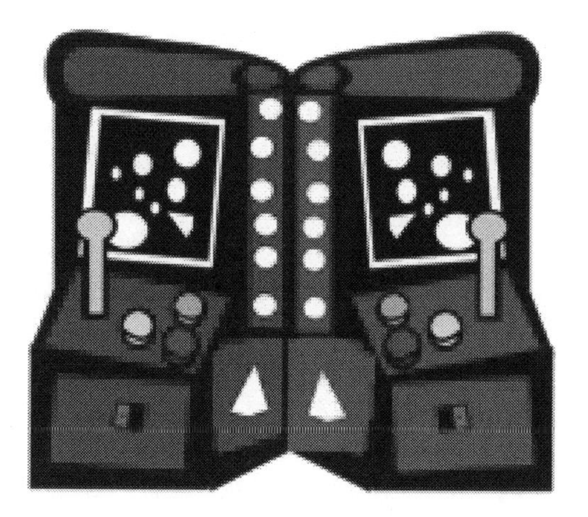

QUIZ GRADE:

A research paper on pinball machines.

- 2-4 pictures of pinball machines 25%
- A concept idea for your pinball machine 50%
- Conclusions and reflections based on your results 25%

TEST GRADE:

Your completed design and the results of the test.

- Project Completed = 50%
- 50% of your grade depends on how long the marble rolls compared to the other group's projects. The projects that do best will get more points.
- *NOTE: There is a -5% penalty for each time the marble gets stuck. You get 1 free restart.*

NOTES:

CATEGORIES: Clothespins, Marbles, Scavengers, Time

MISSION: Running Uphill

BRIEF: You and your team have been selected to make a device that will move water from one cup up to another higher cup.

MISSION RULES:

1. Your design may be of any dimensions less than 18 inches in any one direction.
2. Your device must incorporate a cup a the beginning and the end. Water must be moved from one cup up to the other.
3. You will work with two or three partners. Teams may not be of more than 4 people.
4. You may use any approved scavenged materials from home and school to build your device.
5. There is a 2 minute time limit. Any water not moved at that point is considered lost.

QUIZ GRADE:

A research paper on pumps and water movement devices

- 2-4 pictures of devices 25%
- A labeled blueprint concept idea for your devices 50%
- Conclusions and reflections based on your results 25%

TEST GRADE:

Your completed design and the results of the test.

- Project Completed = 50%
- 50% of your grade depends on how fast you can move the water from one cup to the other. Faster projects score better.

NOTES:

CATEGORIES: Cups, Scavengers, Time, Water

MISSION: Salty Bridges

BRIEF: You and your team have been selected to make the longest bridge as possible from pretzel sticks and glue.

MISSION RULES:

1. You will research bridges and get ideas for a concept for your bridge design.
2. Your bridge must be as long as possible. It may be any height or width required to accomplish this.
3. You will work with two or three partners. Teams may not be of more than 4 people.
4. You must only use thin pretzel sticks and glue for your project. You teacher will determine how many materials you may use.
5. The bridge must not be attached to any surface, but must have a place on each side where it can be rested on a surface, so it can set up to span between two tables, chairs, or something along those lines.

TEACHER'S NOTES: Wax paper is advised for a surface on which to dry the glue and pretzel sticks as it is being assembled.

QUIZ GRADE:

A research paper on bridges.

- 2-4 pictures of bridges 25%
- A concept idea based on your bridge pictures 50%
- Conclusions and reflections based on your results 25%

TEST GRADE:

Your completed design and the results of the test.

- Project Completed = 50%
- 50% of your grade depends on how long your project is compared to the other group's projects. The projects that do best will get more points.

NOTES:

CATEGORIES: Bridges, Glue, Length, Pretzels

MISSION: Scavenger Bikes

BRIEF: You and your team have been selected to make a working miniature version of a bicycle from scavenged materials.

MISSION RULES:

1. You will design a working bike or tricycle. It must have a chain drive or belt drive to allow the pedals to move the rear wheel.
2. The bike may be built from any approved materials found at home or at school. Suggested materials are: cardboard tubes, plastic straws, cardboard, rubber bands, paperclips, glue, tape, etc...
3. You will work with one or two partners. Teams may be of no more than 3 people.
4. Your device may be of any dimensions less than 18 inches.

TEACHER'S NOTES: Style and Function awards are a cool option.

QUIZ GRADE:

A research paper on bikes.

- 2-4 pictures of bikes 25%
- A labeled concept idea based on your bike pictures, including what materials you hope to use for each piece 50%
- Conclusions and reflections based on your results 25%

TEST GRADE:

Your completed design and the results of the test.

- Project Completed = 50%
- 50% of your grade depends on whether or not the belt/chain drive works for the rear tire.

NOTE: Adding brakes or other details may help your score.

NOTES:

CATEGORIES: Bikes, Gears, Rubber Bands, Scavengers

MISSION: Ski Jump

BRIEF: You and your team have been selected to make a ramp to jump marbles, golf balls, and/or ping pong balls as far as possible.

MISSION RULES:

1. You will design a jump to help a ball (marbles/ping pong balls/golf balls) as far as possible.
2. Distance is only measured to the first bounce. Rolling distance is not counted.
3. Your finished project must be built of cardboard tubes, tape, glue, and other materials approved by your teacher.
4. You will work with one or two partners. Teams may be of no more than 3 people.
5. Your project must be freestanding, and may not be attached to any surface.

TEACHER'S NOTES: You might want to have students start collecting cardboard tubes from paper towels and gift wrapping paper a good deal ahead of time before they start the project. Toilet paper rolls are likely not sanitary...

QUIZ GRADE:

Create a blueprint design for your ideas

- Sketch 25%
- Sketch is labeled 25%
- Explanation of strategies 25%
- Conclusions and reflections based on your results 25%

TEST GRADE:

Your completed design and the results of the test.

- Project Completed = 50%
- 50% of your grade depends on how far the ball flies before hitting the ground compared to the other group's projects. The projects that do best will get more points.

NOTE: If you are testing more than 1 type of ball in your design, high averages or best distance per type may receive extra points.

NOTES:

CATEGORIES: Cardboard Tubes, Distance, Golf Balls, Marbles, Ping Pong Balls

MISSION: Slow Rollers

BRIEF: You and your team have been selected to make the slowest ramp possible.

MISSION RULES:

1. You will design a working bike or tricycle. It must have a chain drive or belt drive to allow the pedals to move the rear wheel.
2. The bike may be built from any approved materials found at home or at school. Suggested materials are: cardboard tubes, plastic straws, cardboard, rubber bands, paperclips, glue, tape, etc...
3. You will work with one or two partners. Teams may be of no more than 3 people.
4. Your device may be of any dimensions less than 18 inches.

TEACHER'S NOTES: Style and Function awards are a cool option.

QUIZ GRADE:

A research paper on bikes.

- 2-4 pictures of bikes 25%
- A labeled concept idea based on your bike pictures, including what materials you hope to use for each piece 50%
- Conclusions and reflections based on your results 25%

TEST GRADE:

Your completed design and the results of the test.

- Project Completed = 50%
- 50% of your grade depends on whether or not the belt/chain drive works for the rear tire.

NOTE: Adding brakes or other details may help your score.

NOTES:

CATEGORIES: Friction, Time

MISSION: Spaghetti Bridges

BRIEF: You and your team have been selected to make as strong of a bridge as possible from pasta and glue!

MISSION RULES:

1. You will research bridges and get ideas for a concept for your bridge design.
2. Your bridge must be 18 inches long, between 2 and 6 inches wide, and 2 to 12 inches tall. If you are outside these measurements by more than 1/2 inch, you will be penalized.
3. You will work with two or three partners. Teams may not be of more than 4 people.
4. You must only use pasta and glue for your project. You teacher will determine how many materials you may use.
5. The bridge must have a place in the center of it where a tray can be attached to hold weight. Projects that hold more weight score better.

QUIZ GRADE:

A research paper on bridges.

- 2-4 pictures of bridges 25%
- A concept idea based on your bridge pictures 50%
- Conclusions and reflections based on your results 25%

TEST GRADE:

Your completed design and the results of the test.

- Project Completed = 50%
- 50% of your grade depends on how much weight your project holds compared to the other group's projects. The projects that do best will get more points.
- *NOTE: There is a -5% penalty for every 1/2 inch your project is out of the specifications.*

NOTES:

CATEGORIES: Bridges, Glue, Materials Strength, Pasta, Weight

MISSION: Starchy Goodness

BRIEF: You and your team have been selected to make a the tallest tower possible from only glue and string!

MISSION RULES:

1. You will design a tower.
2. Your device may be any dimensions, but it must be as tall as possible. It must have its own base.
3. You will work with one or two partners. Teams may not be of more than 3 people.
4. You may use only string/yarn and glue to build your project. Your teacher will determine your maximum amount of materials.
5. The tower may not be braced against or attached to any other objects, except for whatever table surface it is built on. Otherwise, it must be completely freestanding.

TEACHER'S NOTES: It is suggested that you use wax paper to set up 'cables' of thread or yarn. Glue will help them set into a specific shape when it dries. Then they can be assembled into a tower.

QUIZ GRADE:

A blueprint design of your idea

- Sketch 25%
- Sketch is labeled 25%
- Explanation of strategy 25%
- Conclusions and reflections based on your results 25%

TEST GRADE:

Your completed design and the results of the test.

- Project Completed = 50%
- 50% of your grade depends on how tall your project is compared to the other group's projects. The projects that do best will get more points.

NOTES:

CATEGORIES: Glue, Height, String, Towers, Yarn

MISSION: Sticky Planes

BRIEF: You and your team have been selected to make as paper airplanes that stick to a wall.

MISSION RULES:

1. You will design a paper airplane that flies across a specified distance and sticks to a target on a wall.
2. The wall may be treated with velcro or tape for stickiness. The design will have to be made in such a way as to hit the wall and stick.
3. No dimension of your airplane should be over 18 inches.
4. You will work with one or two partners. Teams may not be of more than 3 people.
5. Your teacher will determine the amount of and type of materials you receive.
6. There will be 3 attempts, possibly at different distances.

TEACHER'S NOTES: Students might be given a piece of the hook side of velcro strip, and there may be a target made of fuzzy felt on the wall. Additional options include flypaper and magnetic strips.

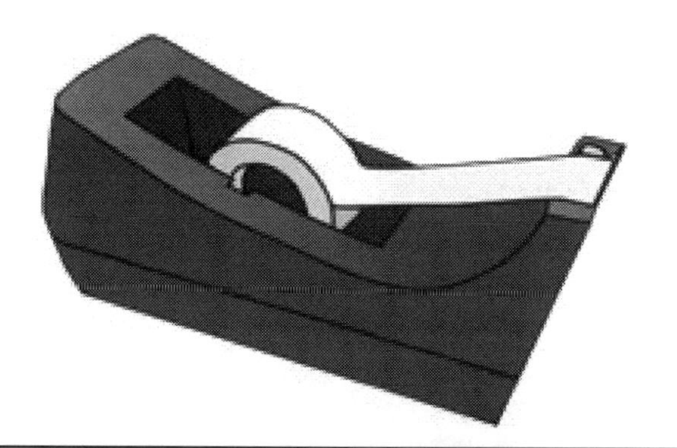

QUIZ GRADE:

A blueprint design of your idea

- Sketch 25%
- Sketch is labeled 25%
- Explanation of strategy 25%
- Conclusions and reflections based on your results 25%

TEST GRADE:

Your completed design and the results of the test.

- Project Completed = 40%
- Project sticks 1 time = +20%
- Project sticks 2 times = +40%
- Project sticks 3 times = +60%

NOTES:

CATEGORIES: Accuracy, Fliers, Paper, Tape, Velcro

MISSION: Straw Rafts

BRIEF: You and your team have been selected to design a raft from plastic straws that can hold the most weight without sinking.

MISSION RULES:

1. You will design a raft with only 10 plastic straws and glue or tape.
2. The raft must float. You will be given 3 chances to see if it floats prior to the real testing.
3. You will work with one or two partners. Teams may be of no more than 3 people.
4. Your boat must have a spot to place a cup or receiver of some sort to hold the weights.
5. After putting your floating boat into the water for the actual test, paperclips will be added until it sinks or tips over and dumps the paperclips.

TEACHER'S NOTE: Pennies, measured amounts of sand, or graduated weights can also be used instead of paperclips.

QUIZ GRADE:

A blueprint design of your idea

- Sketch 25%
- Sketch is labeled 25%
- Explanation of strategy 25%
- Conclusions and reflections based on your results 25%

TEST GRADE:

Your completed design and the results of the test.

- Project Completed = 50% for successfully floating.
- 50% of your grade depends on how much weight the boat holds. The more it holds in comparison to the other teams, the better you do.
- You will be awarded these points in 10% increments. The boat that holds the most clips automatically gets 100%

NOTES:

CATEGORIES: Boats, Buoyancy, Plastic Straws, Water, Weight

MISSION: Stringy Situation

BRIEF: You and your team have been selected to make as strong of a bridge as possible from string or yarn and glue!

MISSION RULES:

1. You will research bridges and get ideas for a concept for your bridge design.
2. Your bridge must be 18 inches long, between 2 and 6 inches wide, and 2 to 12 inches tall. If you are outside these measurements by more than 1/2 inch, you will be penalized.
3. You will work with two or three partners. Teams may not be of more than 4 people.
4. You must only use string (or yarn) and glue for your project. Your teacher will determine the amount of materials you receive.
5. The bridge must have a place in the center of it where a tray can be attached to hold weight. Projects that hold more weight score better.

TEACHER'S NOTES: It is suggested that you use wax paper to set up 'cables' of thread or yarn. Glue will help them set into a specific shape when it dries. Then they can be layered and attached to form the bridge.

QUIZ GRADE:

A research paper on bridges.

- 2-4 pictures of bridges 25%
- A concept idea based on your bridge pictures 50%
- Conclusions and reflections based on your results 25%

TEST GRADE:

Your completed design and the results of the test.

- Project Completed = 50%
- 50% of your grade depends on how much weight your project holds compared to the other group's projects. The projects that do best will get more points.
- *NOTE: There is a -5% penalty for every 1/2 inch your project is out of the specifications.*

NOTES:

CATEGORIES: Bridges, Glue, Materials Strength, String, Weight, Yarn

MISSION: Suspension Bridges

BRIEF: You and your team have been selected to make as strong of a suspension bridge as possible!

MISSION RULES:

1. You will research suspension bridges and get ideas for a concept for your bridge design.
2. Your bridge must be 18 inches long, between 2 and 6 inches wide, and 2 to 12 inches tall. If you are outside these measurements by more than 1/2 inch, you will be penalized.
3. You will work with two or three partners. Teams may not be of more than 4 people.
4. You may use any approved materials from home or school.
5. The bridge must have a string or line running from one side to the other. A cup will be hung from the line in the center of the bridge, where weight will be added.

TEACHER'S NOTES: Paperclips or coins make great weights. The cup will need some sort of hooks, probably paperclips, to attach it to the suspension lines. You may wish to suggest towers/bracing of some sort on the designs.

QUIZ GRADE:

A research paper on suspension bridges.

- 2-4 pictures of bridges 25%
- A concept idea based on your bridge pictures 50%
- Conclusions and reflections based on your results 25%

TEST GRADE:

Your completed design and the results of the test.

- Project Completed = 50%
- 50% of your grade depends on how much weight your project holds compared to the other group's projects. The projects that do best will get more points.
- *NOTE: There is a -5% penalty for every 1/2 inch your project is out of the specifications.*

NOTES:

CATEGORIES: Bridges, Materials Strength, Scavengers, String, Weight

MISSION: Target Practice

BRIEF: You and your team have been selected to make a device that shoots an arrow as accurately as possible.

MISSION RULES:

1. You will design a device that shoots a blunt or **harmless** arrow through a target ring as accurately as possible. Your device should use rubber bands as your major method of propulsion.

2. Your device may be of any dimensions under 9 inches in any one direction.

3. You may build your device from any approved materials found at school or at home.

4. The device must be freestanding and not attached to any surface.

5. You will work with one or two partners. Teams may be of no more than 3 people.

6. At least to three tests will be made from different distances. Success is measured by passing through the target ring.

TEACHER'S NOTES: In testing for accuracy, you can test distance, or test by making progressively smaller target rings to shoot through and keeping the distance the same. 3 attempts at each range are suggested.

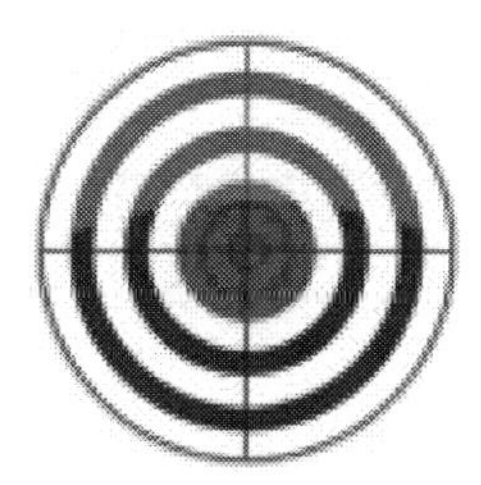

QUIZ GRADE:

A blueprint design of your idea

- Sketch 25%
- Sketch is labeled 25%
- Explanation of strategy 25%
- Conclusions and reflections based on your results 25%

TEST GRADE:

Your completed design and the results of the test.

- Project Completed = 50%
- 50% of your grade depends on how accurately your project can put the arrow through the target ring.
- *NOTE: The best project gets an automatic 100%.*

NOTES:

CATEGORIES: Accuracy, Distance, Rubber Bands, Scavengers, Throwers

MISSION: Throwing Money Away

BRIEF: You and your team have been selected to make a device that throws a coin as far as possible using a clothespin.

MISSION RULES:

1. You will design a device that throws a coin as far as possible, using a clothespin as your spring mechanism.
2. Your primary building materials should be a clothespin and a small cup or holder to hold the coin. You also need a base to hold the entire device.
3. The clothespin must be attached to some sort of base or platform.
4. You will work with one or two partners. Teams may be of no more than 3 people.
5. Up to three tests will be made.

TEACHER'S NOTES: After 3 tests, honors can be given for the best average and/or the highest single shot.

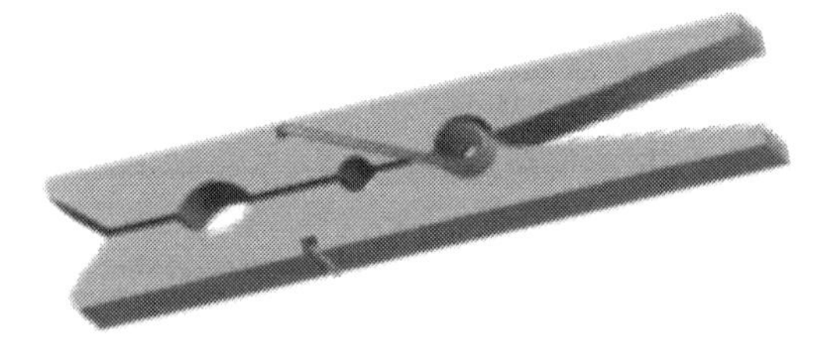

QUIZ GRADE:

A blueprint design of your idea

- Sketch 25%
- Sketch is labeled 25%
- Explanation of strategy 25%
- Conclusions and reflections based on your results 25%

TEST GRADE:

Your completed design and the results of the test.

- Project Completed = 50%
- 50% of your grade depends on how far your project sends the coin.
- *NOTE: The best project gets an automatic 100%.*

NOTES:

CATEGORIES: Clothespins, Coins, Distance, Throwers

MISSION: Trebuchets

BRIEF: You and your team have been selected to make a device that throws a marble as far as possible using a lever and a counterweight.

MISSION RULES:

1. You will design a device that throws a marble as far as possible, using a lever and a counterweight as your spring mechanism.
2. Your device may be of any dimensions under 12 inches in any one direction.
3. You may build your device from any approved materials found at school or at home.
4. The device must be freestanding and not attached to any surface.
5. You will work with one or two partners. Teams may be of no more than 3 people.
6. Up to three tests will be made. Distance will be measured at the first bounce.

TEACHER'S NOTES: After 3 tests, honors can be given for the best average and/or the highest single shot.

Shooting into a long sand pit allows the ball to stop, rather than bounce. Golf balls could be used, too.

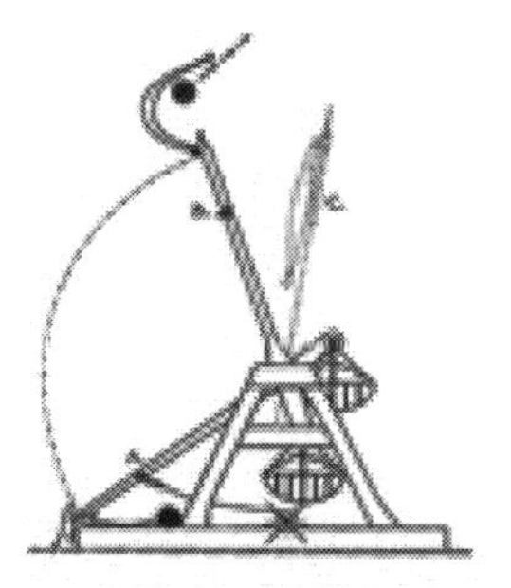

QUIZ GRADE:

A blueprint design of your idea

- Sketch 25%
- Sketch is labeled 25%
- Explanation of strategy 25%
- Conclusions and reflections based on your results 25%

TEST GRADE:

Your completed design and the results of the test.

- Project Completed = 50%
- 50% of your grade depends on how far your project sends the marble.
- *NOTE: The best project gets an automatic 100%.*

NOTES:

CATEGORIES: Distance, Levers, Marbles, Scavengers, Throwers

MISSION: Versus I - The Takedown

BRIEF: One team must design a tower. The other team must knock it down.

MISSION RULES:

1. Each team will consist of 2-4 members.
2. A-Teams will design towers. B-Teams will design machines to knock them down.
3. Each team may design a device from any approved materials found at home or at school.
4. A-Teams and B-Teams should design and build their devices separately.
5. Devices should not be more than 18 inches in any one dimension, and no more than 12 inches in the other dimensions.
6. When completed, A-Teams' towers must survive 5 attacks from B-Teams' devices.

TEACHER'S NOTES: Choosing A-Teams and B-Teams may be done randomly.

You may also make all students be A's or B's, while you take the other job.

QUIZ GRADE:

A blueprint design of your idea

- Sketch 25%
- Sketch is labeled 25%
- Explanation of strategy 25%
- Conclusions and reflections based on your results 25%

TEST GRADE:

Your completed design and the results of the test.

- Project Completed = 50%
- 50% of your grade depends on how much well your project does.
  - A-TEAMS: Project survives with no damage 50%, minimal damage 40%, some damage 30%, completely wrecked, 20%
  - B-TEAMS: Project inflicts complete destruction 50%, mostly complete destruction 40%, some destruction 30%, minimal destruction 20%

CATEGORIES: Crashes, Scavengers, Survival, Throwers, Towers, Versus

MISSION: Versus II - The Splashdown

BRIEF: One team must design a boat. The other team must destroy or sink it.

MISSION RULES:

1. Each team will consist of 2-4 members.
2. A-Teams will design boats. B-Teams will design machines to sink or break them.
3. Each team may design a device from any approved materials found at home or at school.
4. A-Teams and B-Teams should design and build their devices separately.
5. A-Team boats should not be more than 9 inches in any dimension. B-Team devices Devices should not be more than 12 inches in any one dimension.
6. When completed, A-Teams' boats must survive 5 attacks from B-Teams' devices.

TEACHER'S NOTES: Choosing A-Teams and B-Teams may be done randomly.

You may also make all students be A's or B's, while you take the other job.

QUIZ GRADE:

A blueprint design of your idea

- Sketch 25%
- Sketch is labeled 25%
- Explanation of strategy 25%
- Conclusions and reflections based on your results 25%

TEST GRADE:

Your completed design and the results of the test.

- Project Completed = 50%
- 50% of your grade depends on how much well your project does.
    - A-TEAMS: Project survives with no damage 50%, minimal damage 40%, some damage 30%, completely wrecked, 20%
    - B-TEAMS: Project inflicts complete destruction 50%, mostly complete destruction 40%, some destruction 30%, minimal destruction 20%

CATEGORIES: Boats, Crashes, Scavengers, Survival, Throwers, Versus, Water

MISSION: Versus III - The Crashdown

BRIEF: One team must design a catching device. The other team must throw a golf ball through it or knock it down.

MISSION RULES:

1. Each team will consist of 2-4 members.
2. A-Teams will design a catching device that gathers golf balls into a cup. B-Teams will design machines to throw golf balls at those devices.
3. Each team may design a device from any approved materials found at home or at school.
4. A-Teams and B-Teams should design and build their devices separately.
5. Either type of device should not be more than 12 inches in any one dimension.
6. When completed, A-Teams' nets must survive and catch as many of the 5 golf ball throws from B-Teams' devices as possible. A ball that misses the device entirely does not count.
7. Success for the catching device is capturing and holding the golf ball in a tank or trap without breaking or dropping the ball.

TEACHER'S NOTES: Choosing A-Teams and B-Teams may be done randomly.

You may also make all students be A's or B's, while you take the other job.

QUIZ GRADE:

A blueprint design of your idea

- Sketch 25%
- Sketch is labeled 25%
- Explanation of strategy 25%
- Conclusions and reflections based on your results 25%

TEST GRADE:

Your completed design and the results of the test.

- Project Completed = 50%
- 50% of your grade depends on how much well your project does.
    - A-TEAMS: Project catches all 5 balls without falling over or getting broken 50%, 4 balls 40%, 3 balls 30%, 2 balls 20%, 1 ball 10%, no balls 0%
    - B-TEAMS: Other project is unable to catch and stop all 5 balls 50%, drops 4 balls 40%, drops 3 balls 30%, drops 2 balls 20%, drops only 1 ball 10%, catches all the balls 0%

CATEGORIES: Crashes, Cups, Golf Balls, Scavengers, Survival, Throwers, Versus

MISSION: Versus IV - The Rundown

BRIEF: One team must design a catching device. The other team must roll a car through it or knock it down.

MISSION RULES:

1. Each team will consist of 2-4 members.
2. A-Teams will design a catching device that can stop a car without being broken or knocked over. B-Teams will design cars to roll at those devices and crash through them.
3. A-Teams should build their devices of packing peanuts and toothpicks.
4. B-Teams should design a car from scavenged materials that can move under its own power.
5. A-Teams and B-Teams should design and build their devices separately.
6. Either type of device should not be more than 12 inches in any one dimension.
7. When completed, A-Teams' nets must survive and catch B-Teams' cars as many times as possible out of 5 trials.
8. Success for the catching device is stopping or holding the car without breaking or being knocked over. If the car misses entirely, redo the test run.

TEACHER'S NOTES: Choosing A-Teams and B-Teams may be done randomly.

You may also make all students be A's or B's, while you take the other job.

QUIZ GRADE:

A blueprint design of your idea

- Sketch 25%
- Sketch is labeled 25%
- Explanation of strategy 25%
- Conclusions and reflections based on your results 25%

TEST GRADE:

Your completed design and the results of the test.

- Project Completed = 50%
- 50% of your grade depends on how much well your project does.
  - A-TEAMS: Project catches all 5 cars without falling over or getting broken 50%, 4 cars 40%, 3 cars 30%, 2 cars 20%, 1 car 10%, no balls 0%
  - B-TEAMS: Other project is broken or fails to stop the car 5 times 50%, 4 times 40%, 3 times 30%, 2 times 20%, 1 time 10%, no times 0%.

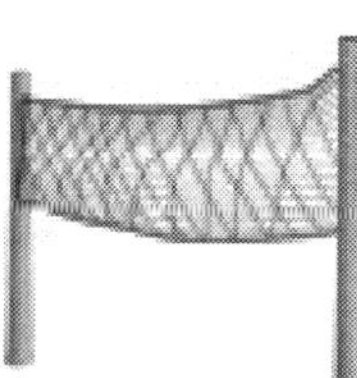

CATEGORIES: Cars, Crashes, Packing Peanuts, Toothpicks, Scavengers, Survival, Versus

MISSION: Volume Up

BRIEF: You and your team have been selected to the largest possible structure possible with only popsicle sticks and rubber bands.

MISSION RULES:

1. Your device may be any possible dimension provided it is made with only the provided amount of rubber bands and popsicle sticks.
2. You will work with one or two partners. Teams may not be of more than 3 people.
3. You must only use rubber bands and popsicle sticks for your project. You teacher will determine how many materials you may use.
4. If the project is roughly rectangular in shape, length x width x height will be used to determine volume, rounding to the nearest inch before multiplying.
5. If the project is more spherical in shape, the formula 4/3 pi r cubed should be used.

QUIZ GRADE:

A blueprint design of your idea

- Sketch 25%
- Sketch is labeled 25%
- Explanation of strategy 25%
- Conclusions and reflections based on your results 25%

TEST GRADE:

Your completed design and the results of the test.

- Project Completed = 50%
- 50% of your grade depends on how large (volume) your project is compared to others.
- The biggest project gets an automatic 100%

NOTES:

CATEGORIES: Popsicle Sticks, Rubber Bands, Volume

MISSION: Washboard

BRIEF: You and your team have been selected to make a device that will squeeze as much water as possible out of a wet sponge.

MISSION RULES:

1. Your design may be of any dimensions less than 18 inches in any one direction.
2. There must be a place to put the wet sponge in your machine, where it can be crushed or squeezed.
3. Your device must incorporate a catch basin to collect water that is squeezed.
4. You will work with two or three partners. Teams may not be of more than 4 people.
5. You may use any approved scavenged materials from home and school to build your device.
6. There must be an activation lever or something that starts your machine.
7. There is a 1 minute time limit. At that time, the test is considered finished.

QUIZ GRADE:

A blueprint design of your idea

- Sketch 25%
- Sketch is labeled 25%
- Explanation of strategy 25%
- Conclusions and reflections based on your results 25%

TEST GRADE:

Your completed design and the results of the test.

- Project Completed = 50%
- 50% of your grade depends on how dry you can get the sponge. Better projects will score higher.

NOTES:

CATEGORIES: Scavengers, Sponges, Time, Water

MISSION: Water Delivery

BRIEF: You and your team have been selected to make a device that will move water from one cup to another.

MISSION RULES:

1. Your design may be of any dimensions less than 18 inches in any one direction.
2. Your device must incorporate a cup a the beginning and the end. Water must be moved from one cup to the other.
3. The first cup cannot simply be poured/ dumped by hand, either.
4. You will work with two or three partners. Teams may not be of more than 4 people.
5. You may use any approved scavenged materials from home and school to build your device.
6. There is a 2 minute time limit. Any water not moved at that point is considered lost.

QUIZ GRADE:

A research paper on pumps and water movement devices

- 2-4 pictures of devices 25%
- A labeled blueprint concept idea for your devices 50%
- Conclusions and reflections based on your results 25%

TEST GRADE:

Your completed design and the results of the test.

- Project Completed = 50%
- 50% of your grade depends on how fast you can move the water from one cup to the other. Faster projects score better.

NOTES:

CATEGORIES: Cups, Scavengers, Time, Water

MISSION: Webbed Up

BRIEF: You and your team have been selected to make a web device that can hold as much weight as possible without breaking.

MISSION RULES:

1. You will design a device using only string or yarn.
2. No single piece of yarn or string may measure more than 12 inches.
3. You may only tie knots to attach the strings. You may not use glue or other adhesives.
4. The device must only be attached at 4-6 anchor points provided by your instructor.
5. You will work with one or two partners. Teams may be of no more than 3 people.
6. Once your device is tied to the anchors, weight will be added gradually to determine how much it can hold before dropping weights or ripping.

TEACHER'S NOTES: C-Clamps, nails in boards, or other anchors should be provided. 4 is a minimum, while 6 (3 per side or in a hex pattern) are suggested.

Suggested weights are: old textbooks, bags of potatoes or similarly-weighted objects. Weights should be large enough to catch in the nets without falling through gaps.

QUIZ GRADE:

A blueprint design of your idea

- Sketch 25%
- Sketch is labeled 25%
- Explanation of strategy 25%
- Conclusions and reflections based on your results 25%

TEST GRADE:

Your completed design and the results of the test.

- Project Completed = 50%
- 50% of your grade depends on how much weight your project holds.
- *NOTE: The best project gets an automatic 100%.*

NOTES:

CATEGORIES: Materials Strength, String, Weight, Yarn

MISSION: What's for Breakfast?

BRIEF: You and your team have been selected to make a device that sorts different shapes and sizes of breakfast cereal into different containers.

MISSION RULES:

1. You will design a sorting device.
2. Your device may be any shape or size, but it must sort the 2 or more types of cereal into different containers, and they must be properly sorted.
3. You will work with one or two partners. Teams may not be of more than 3 people.
4. You may use any approved materials you can find at school or at home, including paper clips, pipe cleaners, plastic straws, notecards, tape...

TEACHER'S NOTES: Use 2 or more different shapes and styles of cereal. Square biscuits, balls, and round rings might be good types to try.

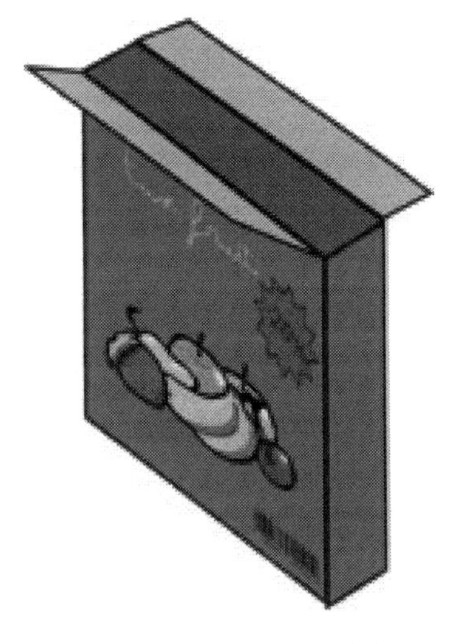

QUIZ GRADE:

Create a blueprint design for your ideas

- Sketch 25%
- Sketch is labeled 25%
- Explanation of strategies 25%
- Conclusions and reflections based on your results 25%

TEST GRADE:

Your completed design and the results of the test.

- Project Completed = 50%
- 50% of your grade depends on the success of your sorting. Each ball is worth 10%, and you must sort 5 balls or marbles.

NOTES:

CATEGORIES: Cereal, Scavengers, Sorting

MISSION: Wooden Cars

BRIEF: You and your team have been selected to make a rubber-band driven wooden car.

MISSION RULES:

1. You will design a rubber-band driven wooden car that uses using spools or other gears to move.
2. The car must use rubber bands as its primary form of propulsion.
3. You will work with two or three partners. Teams may be of no more than 4 people.
4. Suggested wooden materials are: wooden thread spools (plastic may be substituted if wooden ones are not available), toothpicks, popsicle sticks, and dowel rods.
5. Up to 3 tests may be made. The car that rolls the farthest using its set number of rubber bands wins!

TEACHER'S NOTES: Wax paper is suggested as a non-sticky surface for glued pieces to dry upon.

QUIZ GRADE:

A blueprint design of your idea

- Sketch 25%
- Sketch is labeled 25%
- Explanation of strategy 25%
- Conclusions and reflections based on your results 25%

TEST GRADE:

Your completed design and the results of the test.

- Project Completed = 50%
- 50% of your grade depends on how far your car moves.
- *NOTE: The car that moves the farthest gets an automatic 100%*

NOTES:

CATEGORIES: Cars, Glue, Popsicle Sticks, Spools, Toothpicks, Wood

MISSION: Wooden Railway

BRIEF: You and your team have been selected to make a train out of wood and glue!

MISSION RULES:

1. You will design a rolling train of at least 3 cars made entirely from wood and glue.
2. The train must run on a popsicle stick railway at least 18 inches long with at least one curve in it.
3. You will work with two or three partners. Teams may be of no more than 4 people.
4. Suggested wooden materials are: wooden thread spools (plastic may be substituted if wooden ones are not available), toothpicks, popsicle sticks, and dowel rods.

TEACHER'S NOTES: Wax paper is suggested as a non-sticky surface for glued pieces to dry upon.

QUIZ GRADE:

A blueprint design of your idea

- Sketch 25%
- Sketch is labeled 25%
- Explanation of strategy 25%
- Conclusions and reflections based on your results 25%

TEST GRADE:

Your completed design and the results of the test.

- Project Completed = 50%
- 50% of your grade depends on how well your train moves.
- *NOTE: There is a 10% penalty for each restart or time the car(s) come off the track.*

NOTES:

CATEGORIES: Glue, Popsicle Sticks, Spools, Toothpicks, Tracks, Trains, Wood

MISSION: Andrew Frinkle

BRIEF: A quick look at the author of this book and the previous volume (which I hope you have!).

ABOUT THE AUTHOR:

1. Over 10 years of experience in the teaching field with a specialization in math and science education in elementary and middle schools.
2. Award Nominated for teacher of the year.
3. Winner of the Karen Pelz Writing Award for short fiction.
4. Author of over 20 novels books in and over 30 educational books.
5. Black Belt in the Korean Sword Martial Art Geomdo.

SNAZZY PHOTO:

CONTACTS DETAILS:

Email me:

underspacewar@yahoo.com

VIsit me:

- www.underspace.org
- www.littlelearninglabs.com
- www.common-core-assessments.com
- www.50STEMLabs.com

NOTES:

Chicago style is the best style of pizza.

Chili Cheese Dogs are pretty awesome, too.

CATEGORIES: Hands-On, Labs, Math, Measurement, Physics, Science, STEM

MISSION: CHECK OUT THE OTHER COOL BOOKS!

Get the whole series on Amazon or download them on TeachersPayTeachers or TeachersNotebook!

Make sure to check out 50 STEM Lab, 50 More STEM Labs, & 50 STEM Labs Cards!

Made in the USA
Charleston, SC
17 October 2016